Synthesis Lectures on Ocean Systems Engineering

Series Editor

Nikolas Xiros, University of New Orleans, New Orleans, USA

The series publishes short books on state-of-the-art research and applications in related and interdependent areas of design, construction, maintenance and operation of marine vessels and structures as well as ocean and oceanic engineering.

Alexander Arnfinn Olsen

Cathodic Protection
of Offshore Structures

 Springer

Alexander Arnfinn Olsen
Southampton, UK

ISSN 2692-4420   ISSN 2692-4471 (electronic)
Synthesis Lectures on Ocean Systems Engineering
ISBN 978-3-031-77580-2   ISBN 978-3-031-77581-9 (eBook)
https://doi.org/10.1007/978-3-031-77581-9

This Springer imprint is published by the registered company Springer Nature Switzerland AG
The registered company address is: Gewerbestrasse 11, 6330 Cham, Switzerland

If disposing of this product, please recycle the paper.

Preface

Cathodic protection, when used in conjunction with protective coatings, is a common method used to protect immersed parts of steel surfaces from corrosion. This text on *Cathodic Protection of Offshore Structures* offers detailed recommendations on cathodic protection for offshore structures. Traditional seagoing vessels dock at regular intervals, but offshore structures such as Floating Production Storage and Offloading structures (FPSOs) are stationary and are in continuous operation for prolonged periods of time. Therefore, the design of the corrosion protection for a 15-year or longer service life of a floating offshore structure requires special consideration. Some Floating Production Installations (FPIs) have hull designs similar to oil tankers, especially those converted from oil tankers. Others are designed and built as floating production platforms. Cathodic protection systems are to consider the structure to be protected as a whole as well as individual components which are attached to the structure.

The information presented in text is intended solely to assist the reader in the methodologies and/or techniques discussed. This text does not and cannot replace the analysis and/or advice of a qualified professional. It is the responsibility of the reader to perform their own assessment and obtain professional advice. Information contained in this text is considered to be pertinent at the time of publication but may be invalidated as a result of subsequent legislation, regulations, standards, methods, and/or more updated information and the reader assumes full responsibility for compliance.

Southampton, UK   Alexander Arnfinn Olsen

Acknowledgements My deepest gratitude and thanks go to the amazing team at BAE Systems and Babcock Marine for their guidance, patience, and support. I also extend my thanks to the editorial team at Springer, and of course to my wife, without whose support and guiding hand this work would not be possible.

This title is published with the kind permission of the American Bureau of Shipping.

Contents

List of Figures

List of Tables

General

1

1.1   Scope

This text draws on offshore industry best practice and design recommendations for corrosion protection of offshore structures using cathodic protection (CP) systems, and provides guidance for cathodic protection design, installation, and maintenance. The basic approach to the cathodic protection design of the offshore structures is to combine cathodic protection with an effective coating system for the underwater surfaces to be protected. The guidance provided in this text can be applied to both coated and bare submerged surfaces. That said, this text mainly covers the cathodic protection of the submerged external surfaces of floating offshore structures, such as barges, jack-ups, semi-submersible platforms, storage tankers, and buoys, which are static during their normal operating condition. Fixed platforms and concrete platforms are not covered in this text. Moreover, this text covers the submerged areas of appurtenances and openings, mooring lines attached to the structure, sea chests and water intakes up to the first valve. The internal surfaces of flooded compartments/ballast tanks are also considered. This text does not cover the cathodic protection of ships, pipelines, subsea casings, subsea manifolds/flow lines/jumpers/umbilicals, risers, and cables. The cathodic protection of ships is covered in *Cathodic Protection of Marine Vessels*. Furthermore, this text does not cover corrosion control of the internal portions of wells, piping, and associated equipment that may be installed on or attached to structures; underwater pipelines and pipeline risers; and safety and environmental protection aspects associated with cathodic protection are all specifically excluded as relevant national or international regulations exclusively apply.

© The Author(s), under exclusive license to Springer Nature Switzerland AG 2025   1
A. A. Olsen, *Cathodic Protection of Offshore Structures*, Synthesis Lectures on Ocean
Systems Engineering, https://doi.org/10.1007/978-3-031-77581-9_1

1.2   Materials

This text considers structures fabricated principally from carbon manganese or low-alloy steels. Hulls primarily made of other materials such as aluminium alloys, stainless steels, or concrete are not covered. Because some parts of the structure may be made of metallic materials other than carbon manganese steels, the cathodic protection systems should be designed to exert complete control over any galvanic coupling and should minimise risks due to hydrogen embrittlement or hydrogen-induced cracking. Cathodic protection can affect the corrosion fatigue properties of the structure. In general, cathodic protection can have following effects on the fatigue:

- Cathodic protection at typical potential values tends to improve the fatigue performance of steel in seawater; and
- The rate of propagation of fatigue cracks may be accelerated in some steels with highly negative cathodic protection potentials. Fatigue crack growth data may be used to establish whether highly negative potentials are significant.

This text does not include any consideration of possible effects of cathodic protection on the fatigue life of the hull.

1.3   Offshore Corrosion Zones

1.3.1   General

The guidance provided in this text is applicable to the whole submerged zone in seawater and saline mud which can normally be found where the floating structure is anchored, moored, or moving. It is also applicable to appurtenances which may be in contact with muds (e.g., chains, spudcans). To simplify corrosion control of structural steel of offshore structures, offshore external structures can typically be divided into three (3) corrosion zones: the Atmospheric Zone, Splash Zone, and Submerged Zone. In addition, a Mud Zone is considered for jacking or self-elevating structures.

1.3.2   Atmospheric Zone

The atmospheric zone of an offshore structure extends upward from the splash zone. It is exposed to sun, wind, spray, and rain. Corrosion in this zone is typically controlled by the application of a protective coating system.

1.3.3 Splash Zone

The splash zone is intermittently wetted by waves, wind-blown water spray, and tidal action (including the tidal zone). In the Gulf of Mexico, the splash zone is typically about 2 m (6 ft). In Cook Inlet, Alaska, this zone is about 9 m (30 ft). The North Sea splash zone can extend to 10 m (33 ft) during winter storms. Methods for controlling corrosion in the splash zone include both coatings and cathodic protection. Cathodic protection as a supplement to protective coatings can be used in a splash zone that is intermittently submerged. Cathodic protection is only effective when the immersion time is sufficiently long enough for the steel to become polarised. Cathodic protection does not provide protection to the portions of the splash zone that are wet due to sea spray or infrequent immersion.

1.3.4 Submerged Zone (External Areas)

The submerged zone (external areas) is the zone below the splash zone and includes any portion of the structure below the mudline. Corrosion control of the submerged zone can be achieved through cathodic protection or cathodic protection in conjunction with coatings.

1.3.5 Submerged Zone (Internal Areas)

The submerged zone (internal areas) are the internal surface areas of tanks and flooded compartments. Corrosion normally is negligible in compartments that are sealed and have no contact with either the atmosphere or seawater. Whenever possible, the design should provide for sealed compartments. Some structural members are flooded and remain flooded for the life of the structure. To prevent internal corrosion, the flooding valves should be closed after flooding to isolate the flooded chambers from the air, and fresh water should be used if possible. In compartments where circulation of seawater is possible, provisions should be made to reduce internal corrosion. Cathodic protection or a combination of cathodic protection and coatings should be used. Cathodic protection is not effective for gas phase parts of tanks and compartments, and these should be coated internally where condensation may occur. The use of impressed current systems in these compartments should be avoided due to the development of toxic and corrosive chlorine gas at the anode and possibly flammable quantities of hydrogen gas at the cathode. In flooded compartments with a source of organic nutrients, bacterial growth may generate organic acids, carbon dioxide (CO_2), and hydrogen sulphide (H_2S) that can cause corrosion. Bacteria-related corrosion can be controlled through the use of internal cathodic protection, coatings, and microbicides.

1.4   Corrosion Control Considerations

1.4.1   Structure Design Considerations

For better corrosion control, the following should be considered:

(1) Tubular members should be used wherever possible for truss structures. Channels back-to-back and I beam are difficult to protect from corrosion and should not be used for construction

(2) All weld joints should be continuous. If lap joints are used, both edges should be continuously welded. Skip and tack welding should not be used. Bolted and riveted fittings should be avoided

(3) Critical locations of structures, their welds, and the heat-affected zone (HAZ) may be stress relieved to reduce the likelihood of fatigue or corrosion fatigue failures

(4) Piping, such as grout lines, well cutting lines, discharge lines, water supply casings, well casing conductors and pipeline risers can cause shielding and interfere with the flow of cathodic protection current if clustered around a large structural member. Piping/lines not needed for continuous operations should be removed or relocated to avoid cathodic protection shielding. A minimum clear spacing of 1.5 diameters of the smaller pipe should be provided

(5) All steel to be protected should have electrical continuity with the cathodic protection system (preferably by welded contact). This electrical continuity should be provided for the lifetime of the structure. Well casing conductors should be electrically connected to the structure; and

(6) Corrosion rates below the mudline are considered to be low. However, mudline corrosion can be significant for structures with long lifetimes, and all mudline members should be properly connected to the cathodic protection system.

1.4.2   Piping System

Piping, valves, submerged pumps, and other special equipment should also be protected from corrosion by cathodic protection system, particularly when different metallic materials are used.

1.4.3   Mooring Lines

For floating systems, there are two basic mooring configurations: catenary mooring and taut mooring. Conventional composite chain and steel wire rope catenary mooring systems are being used successfully on floating systems in water depths up to 1,000 m (3,280 ft).

However, in depths beyond 1,000 m (3,280 ft), semi-taut and inverted catenary mooring arrangements introduce buoyancy into the mooring line combination to reduce the vertical load of the mooring line. The corrosion protection of all components of the mooring lines should be considered during the design phase so that the moorings are not compromised by corrosion damage. The components include fairleads, chain connectors, chain, and wire ropes. It should be noted that various mixed metallurgy can result in localised galvanic corrosion. The cathodic protection current draining through mooring lines may need additional anode weight for the cathodic protection systems on the hull to which the mooring lines are attached. For mooring lines and anchors, bacterial corrosion in the bottom parts exposed to seabed sediments is to be evaluated with consideration for the use of cathodic protection, coating, or corrosion allowance.

1.4.3.1 Chains

A corrosion allowance is provided for long-term service of chains. A higher corrosion allowance should be provided in the splash zone. The wear could be substantial due to chain handling. Many permanent floating production units require minimal handling of the mooring systems. It has been shown from experience that cathodic protection for the hull structure is also effective for chains extending about 30–60 m (100–200 ft) from the structure. The length depends on the chain connection to the structure, the size of the chain, and the line tension. Galvanic coatings or protective coatings are also used for corrosion protection and fatigue improvement of the chains.

1.4.3.2 Wire Ropes

The large diameter spiral strand rope commonly used on offshore structures normally has its own corrosion protection scheme. The scheme may include blocking compounds, sheaths, sacrificial anode strands and galvanic coatings on the strands. Wire ropes made from galvanised wires, with an outer extruded corrosion protection jacket of polyurethane, polyethylene, or a similar material, are usually used. Grease injection is used for lubrication and to prevent water entrapment. Sacrificial zinc wires may be included. Hydrogen effects should be evaluated as part of any cathodic protection of the ropes due to the extremely high strength of the wires used. It is important to confirm that there is adequate anode material and a superior quality coating on the wire rope connectors.

1.5 Personnel

Personnel responsible for the design, testing, measurements, monitoring, supervision of installation, supervision of operation, supervision of maintenance, and inspection of cathodic protection systems are to have the necessary experience and qualifications in cathodic protection to competently perform their tasks. Personnel are to be trained and certified to achieve and demonstrate the necessary competence levels for the tasks under-taken. EN 15257 provides a suitable method of assessing and certifying the competence of cathodic protection personnel.

1.6   Normative References

The following documents, in whole or in part, are normatively referenced in this text:

EN 12473: General principles of cathodic protection in seawater
EN 12496: Galvanic anodes for cathodic protection in seawater and saline mud
EN 13509: Cathodic protection measurement techniques
EN 15257: Cathodic Protection—Competence levels and certification of cathodic protection personnel
EN 50162: Protection against corrosion by stray current from direct current systems.
EN 16222: Cathodic protection of ship hulls
EN 13173: Cathodic protection for steel offshore floating structures.

1.7   Definitions and Acronyms

1.7.1   Definitions

The following terms and definitions are used in this document:

Anode: The corroding part of an electrochemical corrosion cell. It can be a sacrificial anode or impressed current anode used in cathodic protection.
Cathode: The non-corroding or protected part of an electrochemical cell.
Cathodic Protection (CP): Cathodically protecting a metal surface from corrosion by using sacrificial anodes or impressed current anodes.
Coating System (protective coating system): The entirety of the multiple layers of coating materials applied in a certain sequence. Each layer serves its own special purpose. The combination of these different layers of coatings or paints ideally leads to the best solution possible to protect the substrate surface from corrosion. Any substrate surface treatment is also considered to be part of the coating system.
Corrosion Rate: The rate, usually in mm/year (MPY), at which the corrosion process proceeds. The corrosion rate is always to be calculated from metal loss on one surface, even when occurring on both sides of a steel plate, etc. (Corrosion rate is not to be confused with "steel thickness reduction rate").

Epoxy: A common resin (binder) type in paints or coatings for marine use. Epoxies are normally of two component types, epoxy resin (A component) chemically reacted with a hardener (B component, e.g., amine), resulting in a relatively hard film.

ISO 8501-1 Sa 2½: An ISO surface cleanliness pictorial standard detailing very thorough blast cleaning. When viewed without magnification, the surface shall be free from visible oil, grease, and dirt and from scale, rust, paint coatings and foreign matter. Any remaining traces of contamination shall show only as slight stains in the form of spots or stripes. Sa 2½ corresponds to NACE Grade No. 2 (near white) and SSPC grade SP 10 (near white).

Shadow Effect: Areas to be protected are not "seen" by the anodes of the cathodic protection system installed. More anodes may be needed for complex structures with consideration of the Shadow Effect.

Underwater Hull: Part of the hull below the light waterline.

1.7.2   Acronyms

AC: Alternating Current

Ag: Silver

AgCl: Silver Chloride

CP: Cathodic Protection

DC: Direct Current

DFT: Dry Film Thickness

EMF: Electromotive Force

FPI: Floating Production Installation

FPSO: Floating Production Storage and Offloading

HAZ: Heat Affected Zone

HISC: Hydrogen-induced Stress Cracking

ICCP: Impressed Current Cathodic Protection
   IMR: Inspection, Maintenance, and Repair

IR Drop: Voltage drop across any resistance, which is the product of current (I) passing through resistance and resistance value (R)

MGPS: Marine Growth Prevention Systems

PWHT: Post Weld Heat Treatment

ROV: Remotely operated vehicle

SACP: Sacrificial Anode Cathodic Protection

Zn: Zinc

Design Criteria and Recommendations

2

2.1 General

A cathodic protection system uses galvanic anodes, an impressed current system, or a combination of both. The objective of cathodic protection is to control corrosion of metallic surfaces which are in contact with electrolytes (such as seawater). The cathodic protection system should provide sufficient and well-distributed currents (or well-distributed anodes) to protect each part of the structure and appurtenances so that the potential of each part of the structure can be polarised to be within the limits given by the protection criteria (refer to this chapter, Sect. 2.4) over the design life. Special consideration should be given to areas such as chains, water intakes, thrusters, and sea chests.

2.2 Design Life

The design life of a cathodic protection system for the submerged hull of a floating offshore structure is normally specified by the owner. Either the whole design life or dry-docking interval(s) should be considered. The design life should further take into account any period of cathodic protection active time prior to commissioning of the protected structures. Maintenance, repair and retrofitting of cathodic protection systems for offshore structures are generally very costly and sometimes impractical. It is therefore normal practice to apply at least the same anode design life as for the protected structure with minimal maintenance/retrofitting requirements. However, in certain circumstances planned retrofitting of sacrificial anodes may be an economically viable alternative to the initial installation of exceptionally large anodes. This alternative should then be planned such that necessary provisions for retrofitting are made during the initial design and fabrication.

A. A. Olsen, *Cathodic Protection of Offshore Structures*, Synthesis Lectures on Ocean Systems Engineering, https://doi.org/10.1007/978-3-031-77581-9_2

2.3   Environment

The design of the cathodic protection systems for the offshore floating structures should consider the anticipated service conditions at the location installed or moving to, such as water salinity, temperature, and ice conditions. Because the potential criteria provided in Chap. 4, Sect. 4.4 is a function of the temperature, dissolved oxygen content, salinity, water velocity, and water conductivity, the seawater properties should be established (refer to EN 12473). Unlike ships, the static conditions are normally considered for the floating offshore structures because the durations when dynamic conditions prevail are generally negligible; even some loop current may also be considered.

2.4   Potential Criteria of Cathodic Protection

2.4.1   Protection Potential

The cathodic protection criteria given in Table 2.1 should be referred to for offshore structures. As indicated in the Table, the accepted criterion for protection of carbon steels or low-alloy steels in aerated seawater is a protection potential of −0.80 V or more negative measured with respect to the Ag/AgCl/seawater reference electrode. In the case of mild steel with active sulphate-reducing bacteria (generally in anaerobic conditions), the potential for protection is −0.90 V (instead of −0.80 V) with respect to Ag/AgCl/seawater reference electrodes. For other circumstances, the potential for corrosion control can be estimated using the Nernst equation. It is important to emphasise that with increasing negative potentials, there may be an adverse effect on fatigue properties and a risk of hydrogen embrittlement of susceptible steels. The polarised potential should not be more negative than −1.10 V (Ag/AgCl (Seawater)) for carbon steels and austenitic stainless steels. Table 2.1 of the also provides negative limits for various materials as a recommendation with further descriptions of the detrimental effects from the cathodic protection, which are summarised as follows for applications in floating offshore structures:

(1) Cathodic protection can cause the formation of hydroxyl ions and hydrogen at the steel surface being protected. The formation of these products may cause disbonding of non-metallic coatings at the metal/coating interface. Coating systems compatible with cathodic protection should be considered
(2) Cathodic protection can eliminate the anti-fouling properties of copper-based alloys in seawater
(3) For subsea fasteners connected to a cathodic protection system, hydrogen should always be considered. It is therefore critical to select appropriate materials for subsea fasteners. Carbon steel fasteners with increasing strength have increasing susceptibility to hydrogen. 350 HV hardness as an upper limit is recommended for subsea

fasteners exposed to cathodic protection. Bolts with strengths up to 720 MPa SMYS are typically considered appropriate for cathodic protection compatibility. Bolts fabricated from AISI 316 stainless steel are typically compatible with galvanic anode cathodic protection

(4) Austenitic stainless steels and nickel-based alloys: Those materials in the solution-annealed condition generally are considered appropriate for cathodic protection compatibility. Moderate cold work of these materials does not cause an HISC issue except for UNS S30200 (AISI 302) and UNS S30400 (AISI 304) stainless steels. Hardened austenitic stainless steels with hardness higher than 350 HV are generally not good candidates to receive cathodic protection

(5) Ferritic and ferritic-pearlitic steels: Materials with a specified minimum yield strength (SMYS) up to 500 MPa have proven compatibility with marine cathodic protection systems. All welds should not have hardness greater than 350 HV

(6) Martensitic carbon and low-alloy steels: Materials with an actual yield strength of about 700 MPa and a hardness about 350 HV could fail under cathodic protection. Untempered martensite with the specified minimum yield strength (SMYS) greater than 500 MPa is especially prone to hydrogen-induced stress cracking (HISC). Welding of materials susceptible to martensite formation should be followed by post weld heat treatment (PWHT) to reduce hardness in the heat-affected zone (HAZ) and residual stresses from welding. All welds should not have hardness greater than 350 HV

(7) Ferritic-austenitic ("duplex") stainless steels: These materials are susceptible to HISC regardless of SMYS or specified maximum hardness. Welding may cause increased HISC susceptibility in the weld and the HAZ. Ferrite content rather than hardness of the welds may be controlled to maximum 60–70%. Forgings are more prone to HISC than wrought materials. Small diameter cold bent pipes, used as production control piping for subsea installations, typically have a good cathodic protection compatibility. However, the design should employ a qualified coating system and prevent local plastic yielding

(8) Copper and aluminium-based alloys are generally immune to HISC, regardless of fabrication modes; and

(9) For high-strength titanium alloys, special qualification testing should be considered due to lack of documentation.

2.4.2 Cathodic Protection Evaluation

2.4.2.1 General

The protection criteria and effectiveness of cathodic protection systems should be confirmed by direct measurement of the potential of the structure. However, visual observations of progressive coating deterioration and/or corrosion are indicators of possible

Table 2.1 Summary of potentials versus Ag/AgCl/seawater reference electrode recommended for the cathodic protection of various metals in seawater (adapted from Cathodic Protection of Ships)

Material	Minimum negative potential volts	Maximum negative potential volts
Iron and steel	−0.80 for aerobic environment	−1.10
	−0.90 for anaerobic environment (with active sulphate reducing bacteria)	−1.10
High-strength steels (yield strength > 690 MPa or hardness > 350 HV)	−0.80	−0.83 to −0.95 [1]
Austenitic stainless steel for aerobic and anaerobic conditions N_{PRE} = % Cr + 3.3% (Mo + 0.5W) + 16% N	−0.30 for $N_{PRE} \geq 40$ [2]	−1.10
	−0.60 NPRE < 40 [2]	−1.10
Duplex stainless steel for aerobic and anaerobic conditions	−0.60 [2]	Refer to note 3
Martensitic stainless steel (13% Cr) for aerobic and anaerobic conditions	−0.50	Refer to note 5
Nickel-based alloys	−0.20	Refer to note 4
Copper alloys	−0.45 to −0.60 for alloys with aluminium	−1.10
	−0.45 to −0.60 for alloys without aluminium	No limit

Notes

(1) For high-strength steels susceptible to hydrogen-induced stress cracking (HISC), the maximum negative potential should be more positive (less negative) than −0.83 V (Ag/AgCl/seawater reference electrode)

(2) For most applications, these potentials are adequate for the protection of crevices, although higher (more negative) potentials may be considered

(3) Forgings, castings, and welds are more prone to HISC than wrought materials due to the coarse microstructure allowing HISC propagating preferentially in the ferritic phase

(4) High-strength nickel copper and nickel chromium iron alloys can be subject to HISC, and potentials that result in significant hydrogen evolution should be avoided

(5) Depending on strength, specific metallurgic condition, and stress level encountered in service, those alloys can be susceptible to hydrogen embrittlement and cracking. If a risk of hydrogen embrittlement exists, then potential more negative than −0.80 V should be avoided

inadequate protection. Steel plate thickness gauging can measure deficiencies in corrosion protection. Storm waves or strong tides can produce high water velocities that tend to depolarise the structure. Depolarisation is less likely to be a problem for well-polarised structures with well-formed calcareous deposits or for coated steel structures. All potential measurements on the structure should be made before removal of marine growth because the removal process could depolarise the steel.

2.4.2.2 Reference Electrode

The Ag/AgCl/seawater reference electrode potential is somewhat affected by the resistivity (caused by the chloride content) of the seawater in which the electrode is immersed. If the resistivity is known to differ appreciably from that of ordinary seawater [20 Ω cm at 20 °C (7.9 Ω in at 68°F)], the electrode reading should be corrected. Other standard reference electrodes, such as high-purity zinc electrodes, may be used to substitute the Ag/AgCl/Seawater. Refer to this chapter, Sect. 3.4 of *Cathodic Protection of Marine Vessels* for more details.

2.4.2.3 Potential Measurements

Potential measurements should be made with the reference electrode and a high impedance (minimum 10 MΩ) voltmeter. The reference electrode should be located in the seawater away from anodes and as close as practicable to the structure to minimise voltage drops. Some measurements should be taken in areas of greatest shielding in evaluating the protective level of a structure. Consideration should be given to voltage (IR) drops other than those across the steel/water interface when structure potential data are evaluated. Changes in water resistivity or temperature variation can affect the voltage drop. In impressed current systems, under conditions involving high-resistivity water or high current density, the voltage drop may be excessive. Potential measurements taken immediately after turning off the rectifier(s), called Instant-Off measurements, may provide useful information by eliminating voltage drop in the water. For galvanic anode protection systems, the included voltage drop is generally not significant in ordinary seawater if the reference electrode is placed close to the structure. The voltage drop may become significant in brackish waters. In such cases, it may be necessary to use interruptible coupons or other IR correction techniques to determine the true potential of the metal surface. The methods of measuring potentials can be involved in diverse ways:

(1) The reference electrode can be suspended freely in the water from a designated location on the structure. The location of the electrode may not be known because of drift resulting from water currents. It may not be appropriate for the identification of problem areas on protected structures. This method is used to assess the general condition of the cathodic protection system

(2) The reference electrode may be run down along a guide wire for control of its location. The guide wire may be permanently installed on the structure, or it may be

temporarily installed with the aid of a heavy weight to anchor the guide wire at the bottom. If the guide wire is metallic, it should be electrically isolated from the structure

(3) The reference electrode can be carried by a diver or a remotely operated vehicle (ROV). This method can identify exact electrode locations with details of a potential survey. Diver safety should be considered during evaluation of impressed current protected structures. At least a part of the ICCP system needs to be de-energised during the cathodic protection measurements. The effect of such de-energisation on the level of protection should be considered during the evaluation of the potential measurements. Diver safety should be considered during evaluation of impressed current protected structures

(4) Some permanent reference electrodes may be mounted on a structure with known electrode locations. The potentials measured from these electrodes are limited to the adjacent structure surfaces of the fixed locations. This method can better track potential changes with time. The accuracy of permanently installed electrodes should be periodically checked against another electrode. Dual element Ag/AgCl/Seawater—Zn electrodes are useful to provide self-calibration

(5) Current density and current output from representative galvanic anodes may also be measured for some structures by installed monitoring systems. Those measurements are particularly useful for new structure designs or under new environments in which precise cathodic protection design criteria are not available; and

(6) By using specially designed reference electrode arrays, the voltage gradient in the seawater around the structure is measured. Although these measurements do not determine the level of protection on the structure, they may be useful in determining current distribution and remaining anode life.

2.4.2.4 Steel Coupons

The corrosion rate should remain within limits acceptable for the intended structure life. Steel coupons, normally small and similar to the structural steel, can be used to determine the effectiveness of corrosion control. Care should be taken in applying the coupon data to other areas or the entire structure on protected structures.

2.5   Coatings in Combination with Cathodic Protection

The cathodic protection design for a floating structure is usually used in combination with a coating system. The use of coatings can drastically reduce the cathodic protection current demand, especially for applications where the current demand for cathodic protection of bare metal surfaces is expected to be high (including deep water applications for which the formation of calcareous deposits may be slow). For weight-sensitive structures with a long design life, the combination of a coating and cathodic protection is likely to give the most cost-effective corrosion control. For some systems with exceptionally long design lives, cathodic protection may be impractical without using coatings. In addition to current demand reduction due to coatings, coatings can also improve current distribution. A consequence of cathodic protection application is the formation of a calcareous layer (consisting primarily of calcium carbonate) on bare metal surfaces to be protected. The thickness of the layer is typically of the order of a tenth of a millimetre, but thicker deposits may occur. The calcareous layer acts as an oxygen diffusion barrier which can reduce the current demand for cathodic protection.

Coatings to be used in combination with cathodic protection should be compatible with cathodic protection. The compatibility is established by a pre-qualification testing of the coating product within applicable cathodic protection potential limit for avoiding cathodic disbonding of the coating. By nature, coatings can deteriorate during their service lives. To compensate for this, the design coating breakdown factors used for cathodic protection design are deliberately selected in a conservative manner so that a sufficient total final current output capacity from the cathodic protection system is maintained. The design coating breakdown factors depend on the type of coating system installed and the service environment experienced. The coating systems used for corrosion protection of the floating offshore structures for 15+ years of service life are typically:

(1) Typical protective coating system for underwater hull
   - Two or more coats of epoxy, polyurethane, or vinyl
   - Total nominal dry film thickness 350 μm (14 mils)
   - A surface cleanliness of ISO Sa 2½ (ISO 8501-1)
   - 75 μm (3 mils) surface roughness
   - A soluble salt limit of 40 mg/m^2 (3.7 mg/ft^2)
(2) Typical protective coating system for flooded compartments/ballast tanks
   - Two coats of epoxy
   - Total nominal dry film thickness 320 μm (12.5 mils)
   - A surface cleanliness of Sa 2½ (ISO 8501-1)
   - 75 μm (3 mils) surface roughness
   - A soluble salt limit of 40 mg/m^2 (3.7 mg/ft^2).

Metallic coatings based on zinc or aluminium are compatible with galvanic anode cathodic protection. However, compared to organic coatings, they generally provide no advantage in decreasing the cathodic protection current demand. Zinc rich primers are considered unsuitable for application with cathodic protection due to either susceptibility to cathodic disbondment or low electrical resistivity, leading to high cathodic protection current demand. For components in materials sensitive to hydrogen induced stress cracking (HISC) by cathodic protection, generally, a non-metallic coating system should always be considered as a barrier to hydrogen adsorption.

2.6   Design Current

2.6.1   General

To achieve the protection potential criteria denoted in this chapter, Sect. 2.4, the appropriate design current density for each part of the structure and appurtenances with respect to the environmental and service conditions should be used for design. The design current of a cathodic protection system should be determined in accordance with the following parameters: structure subdivision, components description and service conditions. Special considerations should be given for areas such as chains/ropes, water intakes, thrusters, and sea chests. Current draining from other parts of an installation (e.g., risers, mooring systems) are to be considered as part of the cathodic protection design. Storm waves or strong tides can produce high water velocities that tend to depolarise the structure. Higher water levels also add additional areas of unprotected steel and increase the current required to produce protective potentials. Depolarisation is less likely to be a problem for well-polarised structures with well-formed calcareous deposits or for coated steel structures.

2.6.2   Cathodic Protection Zones

In the design of cathodic protection systems, the submerged part of a floating structure can be divided into different cathodic protection zones (cathodic protection zones) based on water depth or physical interfaces of the protected structure or surface temperature (thus affecting current density). Those zones are considered independently with respect to cathodic protection design, although they may not necessarily be electrically isolated from each other. Some specific components may constitute a cathodic protection zone, such as thrusters, rudders, propellers, and openings of sea chests. For a storage tanker, sea chests are considered as separate cathodic protection zones apart from the underwater hull cathodic protection zone. The aft cathodic protection zone may include the aft part of the hull, propellers, shafts, and rudders. For buoys, the body of the buoy is considered a zone and the influenced part of the mooring chain(s) is considered as another zone.

The cathodic protection system is determined and dedicated for each cathodic protection zone. Each component should be fully detailed in the design including material, surface area and coating characteristics (coating type, design life and coating breakdown factor). Special considerations should be given for areas such as water intakes, thrusters, and sea chests:

(1) Electrochemical marine growth prevention systems (MGPS) are often used within sea chests to prevent fouling of seawater intake systems and may interact with

the cathodic protection system. This effect should be considered in the design and installation of the anti-fouling system; and

(2) Cathodic protection within sea chests may adversely affect box coolers in sea chests if the box coolers are electrically isolated from the sea chest. This possible effect should be taken into account in designing the cathodic protection requirements for sea chests.

The cathodic protection current density is strongly dependent on water temperature and depth. For deepwater structures, distinctive design values should be used for different temperature zones. To optimise the design, the structure should be divided into separate zones wherein the temperature does not vary by more than 5 °C (9°F). The average temperature of each interval should be used to assess the required current densities. The same approach can apply to resistivity assessment.

2.6.3 Surface Area Calculation

For each cathodic protection zone, surface areas to receive cathodic protection should be calculated separately for surfaces with and without a coating system. Surface areas affected by other parameters (e.g., high, and very cold surface temperatures) need special consideration with respect to the influence on the cathodic protection current demand. For internal surfaces, complex geometries sometimes exist within tanks, such as stiffeners and heating coils. Lower sections of tanks not fully drained within stiffeners can also constitute discrete cathodic protection zones to be considered. Each component of a cathodic protection zone should be fully detailed in the design. This should include:

(1) Material type
(2) Specific potential limit (if applicable)
(3) Complexity of the structure
(4) Surface area; and
(5) Coating characteristics, including type, predicted lifetime, anticipated coating breakdown.

It is acceptable to simplify calculations of the surface areas of complex geometries, provided that the simplification is conservative.

2.6.4 Current Demand

2.6.4.1 General

The current density required may not be the same for all components of the structure, as the environmental and service conditions are variable. The selection of design current densities may be based on experience gained from similar structures in a similar environment or from specific tests and measurements. Current density depends on the kinetics of electrochemical reactions and varies due to factors such as the protection potential, surface condition, dissolved oxygen content in seawater, seawater velocity at the steel surface and temperature. The following should be evaluated for each design:

(1) Initial current density, i_{ci}, required to achieve the initial polarisation of the structure. The initial structure surface polarisation is normally short compared to the design life. A calcareous deposit can be formed during the initial polarisation from high current density applied, which affects the maintenance current density (mean current density) requirement

(2) Mean current density, i_{cm}, which is also called as the average or maintenance current density, required to maintain polarisation of the structure. The mean current density is used to calculate the minimum mass of anode material necessary to maintain cathodic protection throughout the design life.

(3) Final current density, i_{cf}, required to protect the metal surface with established marine growth and calcareous layers, which takes into account the current density required for possible repolarisation of the structure (e.g., after severe storms or cleaning operations) and continuing coating breakdown (where coated).

Typically, the number of anodes required for protection should satisfy three different criteria. There should be enough anodes to (a) polarise the structure initially (initial current density), (b) produce the appropriate amount of amps of current over the design life of the structure (mean current density), and (c) produce enough current to maintain protection at the end of the design life (final current density).

2.6.4.2 Current Density

The current density requirements for the underwater hull of floating offshore structures are strongly influenced by the specific environmental conditions where the structure operates. Recommended initial design current densities (mA/m^2) for seawater exposed bare metal surfaces to be polarised, as a function of depth and 'climatic region' based on surface water temperature, are given in Table 2.2.

Generally, mean current density is about half of the initial current density (i.e., $i_{cm} = \frac{1}{2}$ of i_{ci}). Final current density is about 2/3 of the initial current density (i.e., $i_{cf} = 2/3$ of i_{ci}). The temperature in the deeper waters will generally always be lower than that in the upper 30 m (100 ft) of water. The initial current density for saline muds around ambient

Table 2.2 Recommended initial design current densities for bare metal surfaces exposed to seawater

Climatic region	Initial current density, i_{ci} (mA/m^2 [mA/ft^2]), as a function of depth and temperature			
	0–30 m (0–100 ft)	>30–100 m (100–328 ft)	>100–300 m (328–984 ft)	>300 m (984 ft)
Tropical region [>20 °C (68°F)]	150 (13.9)	120 (11.1)	140 (13.0)	180 (16.7)
Sub-tropical region [>12~20 °C (>54~68°F)]	170 (15.8)	140 (13.0)	160 (14.9)	200 (18.6)
Temperate region [>7~12 °C (>45~54°F)]	200 (18.9)	170 (15.8)	190 (17.7)	220 (20.4)
Arctic region [≤7 °C (≤45°F)]	250 (23.2)	200 (18.6)	220 (20.4)	220 (20.4)

temperature can be 25 mA/m^2 (2.3 mA/ft^2). The mean and final current density can be 20 mA/m^2 (1.9 mA/ft^2). The current density depends on seawater temperature, oxygen content and salinity. It is known that with increasing water depth:

- Salinity may increase or decrease depending upon geographical location
- Oxygen concentration decreases with depth, but remains significant
- Seawater temperatures decrease
- Calcareous deposits become increasingly more soluble and less persistent; and
- Resistivity typically increases.

2.6.4.3 Current Demand

The current demand of each metallic component of the structure is the result of the product of its surface area multiplied by the required current density.

$$I_c = A_c \times i_c \times f_c$$

The equation applies to initial, mean, and final current demand calculations. The "current density," i_c, refers to cathodic protection current per unit surface area. The "coating breakdown factor," f_c, affects current demand. The initial current demand (I_{ci}) is extremely low and not critical for a fully coated structure. It does not need to be part of the cathodic protection design. However, calculation of the initial current demand should be included in a cathodic protection design report for future reference. The cathodic protection current demand calculations should include both the mean current demand (I_{cm}) and the final current demand (I_{cf}) calculations. If the underwater hull retrofitting of anodes is planned, no final current demand (I_{cf}) calculation is needed. The cathodic protection current demand from anodes meets following condition:

$$I_{total\ mean} \geq I_{cm}$$

$$I_{total\ final} \geq I_{cf}$$

2.6.4.4 Coating Breakdown Factor, f$_c$, for Coated Steel

The usual approach to a cathodic protection design for a floating structure is a combination of cathodic protection and a coating system. The recommended offshore structure coating systems are to refer to this chapter, Sect. 2.5. Following formula is provided to estimate coating breakdown factors:

$$f_c = \alpha + \beta t$$

where

α initial coating breakdown,
β coating deterioration rate annually.
t service years.

Parameters α and β fully depend on the coating system used, coating application and in-service maintenance/repair of the coating. If there is no other data available, for typical two-coat coating systems recommended in this chapter, Sect. 2.5, initial coating breakdown $\alpha = 0.01$ (i.e., 1%) is assumed and the following f_c can be used. For underwater hull depth up to 30 m (100 ft):

- The initial coating breakdown factor, $f_{ci} = 0.01$
- The mean coating breakdown factor, $f_{cm} = 0.01 + 0.006t$
- The final coating breakdown factor, $f_{cf} = 0.01 + 0.012t$.

For underwater hull depth greater than 30 m (100 ft):

- The initial coating breakdown factor, $f_{ci} = 0.01$
- The mean coating breakdown factor, $f_{cm} = 0.01 + 0.004t$
- The final coating breakdown factor, $f_{cf} = 0.01 + 0.008t$.

When a one-coat coating system with a reduced coating standard compared to that of the recommended coating systems in this chapter, Sect. 2.5 is used, the α and β values should be increased to estimate the coating breakdown factors.

2.7 Circuit Resistance

2.7.1 General

The number and location of the anodes determines an electrical current distribution for achieving the protection potential level over the whole steel structure surface. The potential drop from cathodic protection circuit resistance should be considered for applied potential to the steel structure surface. For cathodic protection systems using galvanic anodes, the optimum anode's size and shape may be determined using Ohm's law of cathodic protection circuit:

$$I = \frac{\Delta E}{R}$$

where

I current output from anode, in amps

ΔE closed circuit driving voltage between anode and structure, in volts

R circuit resistance, which is sum of anode-to-electrolyte resistance, electrolyte resistance and structure-to-electrolyte resistance, in ohms.

The circuit resistance, R, is assumed to be equal to the anode/electrolyte resistance or anode resistance as the structure-to-electrolyte resistance is exceedingly small in seawater. ΔE is generally taken as the potential difference between the polarised potential of the steel and the operating potential of the particular anode alloy in seawater (i.e., closed circuit voltage). For an impressed current system, the DC output voltage of the power source should be higher than the sum of the voltage drops in all the components of the circuit cables, electrolyte resistance, the anode-to-electrolyte resistance and structure-to-electrolyte resistance, and back EMF between the anode and steel structure. The voltage between the ICCP anode and electrolyte should not exceed the maximum acceptable value depending on the material of the anode.

2.7.2 Anode Resistance Calculations

The electrical resistance of an anode to the surrounding electrolyte depends upon electrolyte resistivity and on the size and shape of the anode. Empirical formulae for the anode resistance calculations can be found in this chapter, Sect. 2.6 of *Cathodic Protection of Ships*. The resistivity value (ρ) of seawater is used for the anode resistance calculations, which should be locally determined. A resistivity value (ρ) of 20–25 Ω cm (7.9–9.8 Ω in) can be used for seawater if a measured value is not available.

Cathodic Protection Systems

3

3.1 General

Cathodic protection of a floating structure may be achieved by either a sacrificial anode cathodic protection (SACP) system or an impressed current cathodic protection (ICCP) system, or both. For the ballast tanks and other tanks containing seawater, an ICCP system should not be used, since it may generate excessive hydrogen and chlorine gas, which may be hazardous. The metals commonly used as sacrificial anodes are aluminium, zinc, and magnesium. These metals are alloyed to improve the long-term performance and dissolution characteristics. In this chapter, Sect. 3.5 provides recommendation on sacrificial anode cathodic protection systems. Impressed-current systems employ inert (exceptionally low dissolution) anodes and use an external source of DC power to impress a current from an external anode onto the metallic surface to be cathodically protected. In this chapter, Sect. 3.6 provides requirements on impressed current cathodic systems.

Cathodic protection systems are to be designed to deliver sufficient current to the structure to be protected for the design life of the structures, so that the selected cathodic protection criteria can be efficiently satisfied for all parts of the structure to be protected. Cathodic protection systems are also to be designed to minimise the effect on associated pipelines/risers/mooring lines and other neighbouring metallic structures. Cathodic protection system design should consider cathodic protection system life extension, when necessary, by providing adequate rehabilitation procedures and appurtenances that may be used to simplify retrofits to the anode system and the impressed current system.

A. A. Olsen, *Cathodic Protection of Offshore Structures*, Synthesis Lectures on Ocean Systems Engineering, https://doi.org/10.1007/978-3-031-77581-9_3

3.2   Cathodic Protection System Selection

3.2.1   Comparison of SACP and ICCP Systems

The decision to use a SACP or an ICCP system, or both, for the structure should be taken at the conceptual stage. To help with selection of cathodic protection systems, a comparison of sacrificial and impressed current anode systems is provided in Table 3.1.

The use of galvanic anodes is appropriate under the following conditions:

- When a relatively small amount of current is required
- Usually when lower-resistivity electrolytes, such as seawater and mud, are present
- Easier to provide local cathodic protection to a specific area on a structure; and
- When additional current is needed at problem areas, such as isolated points from over-all impressed current cathodic protection systems or for electrically shielded areas caused by non-uniform current distribution from remotely located impressed current systems.

Impressed current cathodic protection systems are used for offshore structures for following conditions:

- When there are large current requirements and weight, and flow resistance are a concern; and
- Operations in waters in which the resistivity changes.

Present experience indicates that for 15+ years of service life the sacrificial anode cathodic protection (SACP) system is considered to be the most effective option for the underwater hull. There are several reasons for this:

- Ability to confidently design for 15+ years of life
- No maintenance required
- Extremely high reliability
- No modifications on hull interior, and no hull penetrations
- Minimal risk of electric current interference
- Compatibility with other cathodic protection systems on subsea equipment; and
- Lower overall life cycle cost for hulls installed in deep water sites.

A SACP system may be required prior to the commissioning of an ICCP system and/or additional sacrificial anodes to dedicated items (e.g., for sea chests).

Table 3.1 Comparison of galvanic anodes and impressed current cathodic protection systems for offshore structures

Comparison item	Sacrificial anode systems	Impressed current systems
Design and installation and maintenance costs	Simple in design and installation, generally no maintenance and supervision required, but costly labour of installation labour and anode replacement when consumed Wrong connection is not possible	Needs careful design and installation. Regular maintenance and monitoring are needed Initial equipment cost is higher, but life cycle cost is lower Wrong connection is possible
Consequence of anode damage	Where a system comprises a large number of anodes, the loss of a few anodes has little overall effect on the system	Loss of anodes can be overly critical to the effectiveness of a system
Environment effect on cathodic protection efficiency	Only practical for low resistivity of electrolyte, such as seawater and mud. Protection potential and current are not controllable	Less restriction from electrolyte resistivity. Protection potential and current can be automatically controlled by ICCP controller
Detriment to coating and steel	Coating system is selected for resisting cathodic disbonding. Low potential anode material is needed for high strength steels	Due to high anode current, the structure can be over polarised and detrimental to coatings and high strength steel if not controlled
Power source	No electric power supply is needed. Can be used where electrical power is not available	Continuous DC power supply is required
Water flow and weight increases	Bulk of anode material may restrict water flow and increase weight/turbulence/noise/drag on the hull Galvanic anodes may interfere with subsea operations and increase drag forces by flowing seawater	Lighter and fewer in number. Anodes may be designed to have minimum effect on water flow Low hull profile reduces noise and drag
Interaction	Less likely to affect any neighbouring structures	Effects on other structures near the anodes need to be assessed

3.2.2   Information for Cathodic Protection System Design

The following information should be considered when the cathodic protection system is selected:

(1) Structure specifications
  • Structure drawings with specifications of materials, coatings, maximum service temperature, water/liquid type, and level
  • Locations of electrical isolation flanges or standoffs
  • Availability of electrical power
  • Safety requirements
  • Installation accessibility
(2) Offshore site conditions
  • Water depth, oxygen content, velocity, turbulence, temperature range with consideration of water depth, water resistivity, tidal and storm effects
  • Adjacent facilities, including pipelines
  • Existing and proposed cathodic protection systems; and
  • Electrical continuity isolation from foreign pipelines or structures.

Complete field survey may be needed if previous experience and test data are not available for estimating current requirements and system performance.

3.3   Electrical Continuity and Current Drain

3.3.1   Cable Connection

If the cathodic protection design utilises cables for electrical continuity, requirements to verify electrical continuity should be specified in the cathodic protection design. It is recommended that the product of the total connection resistance in the circuit and the current demand (or current output for a non-welded anode) does not exceed 10% of the design driving voltage. In no case should the resistance across a continuity cable exceed $0.1\ \Omega$.

3.3.2   Electrical Bonding

When cathodic protection is required for metallic appurtenances such as rudders, propellers, turrets, swivel fairleads and thrusters, electrical bonding of the metallic appurtenances to the hull structure should be provided by appropriate means unless the appurtenances are protected by independent cathodic protection systems. This electrical bonding with low resistance is to be maintained to provide adequate cathodic protection of the appurtenances connected.

(1) To prevent galvanic corrosion of the hull or bearings, it is necessary to bond corrosion-resistant copper-based alloys or stainless-steel propellers or thrusters to the adjacent hull
(2) Rudders and turrets should be bonded by means of flexible cables connected to the adjacent hull generally by welded/brazed studs. Allowances should be made for rudder movement by providing a large loop of ground strap; and
(3) For buoys and other moored structures, no particular continuity device with anchor chains is generally required but continuity should be assessed.

Cable connections to the hull should be of a welded or brazed type or threaded connections without coating. Coatings on contact surfaces should be removed prior to assembly. If the contact is made by using copper cables welded or brazed at each end, these cables should be stranded and have a minimum cross-section of 16 mm^2 (0.025 in^2). If cable shoes are used, the copper cable should be brazed to the cable shoe. For hulls equipped with a propulsion system, a turning propeller shaft should be electrically insulated from the hull by the lubricating oil film in the bearings and by the use of non-metallic bearing materials in the tail shaft. When the shaft is insulated in this way, an electrical potential can be measured between the shaft and the hull, and this can cause corrosion. The effectiveness of the shaft ground assembly system should provide a maximum contact resistance of no greater than 0.001 ohms for a water filled bearing and 0.01 ohms for an oil filled bearing. The potential readings through mV meter should be checked and maintained below a level. A maximum value of 50 mV is recommended unless otherwise specified.

3.3.3   Connection to Other Structures

A floating structure may be permanently or temporarily connected to other neighbouring structures. Each structure should be fitted with its own cathodic protection system which should be checked before electrically connecting it to the floating structure under consideration. If foreign structures are not fitted with a cathodic protection system and are temporarily connected to the protected structure, the potential of the protected structure should be measured to confirm that the protection is being maintained at an acceptable level during the period of connection. Possible detrimental effects on the operation of the cathodic protection system should be evaluated. All components or structures intended to be electrically connected to the cathodic protection system are to be considered in the cathodic protection current drain calculations, and include but are not to be limited to the following:

- Connectors/risers
- Mooring systems
- Turrets and conduits
- Structural appurtenances; and
- Any electrically connected components which are fully resistant to corrosion (e.g., items made of corrosion resistant alloys).

It is recommended that interaction testing should be carried out to demonstrate that adjacent structures are not adversely affected by the new cathodic protection system with the levels permitted in EN 50162:

- Adjacent structures fitted with cathodic protection should not have their protection levels changed beyond the levels indicated in EN 50162
- Adjacent structures not fitted with cathodic protection should not have their corrosion potentials changed by more than +20 mV by the new cathodic protection system as defined in EN 50162; and
- Similarly, it is recommended that cathodic protection interaction testing be performed to determine the effects on the vessel's hull/structure by the adjacent cathodic protection system.

3.4   Stray Current

Stray electrical currents are undesirable electrical current flows. When a protected structure lies near other immersed or buried metallic structures, the metallic structure may pick up a proportion of the protective current due to potential gradients in the water or mud and return it to the water/mud at other points where corrosion will take place. On an FPSO, this risk may exist on mooring components close to the hull, or on mechanical couplings associated with the risers within the turret structure. The method for overcoming this effect depends on the circumstances but may include the use of drainage bonds, by insulating flanges or joints, or by insulating structures with suitable coatings. It is important that all electrical equipment on the vessel/structure be installed in a manner which can avoid stray current in the hull. In the case of the installation of an ICCP system for the underwater hull all electrical wiring and connections are to be checked to avoid stray currents.

Monitoring the underwater hull early in the service life of the vessel can confirm that stray current corrosion does not occur on the hull. If the measured potentials show relatively constant values in the range −900 to −1000 mV (reference electrode Ag/AgCl/ Seawater) it indicates the absence of stray current corrosion. A measured local peak in the potential (e.g., a measured potential more positive than about −800 mV) would indicate a possible stray current corrosion situation on the hull and that further action is needed.

3.5 Galvanic Anode Cathodic Protection

3.5.1 General

A galvanic (also called sacrificial) anode is a metal that has a more negative potential than the structure metal to be protected. Galvanic anodes corrode more readily than the protected structure, providing protection to the structure. Galvanic anodes for marine applications are usually made of zinc- or aluminium-based alloys. Magnesium-based alloy anodes can be used for freshwater applications, but not for seawater application because of their high potential and high conductivity of seawater. The objective of a galvanic anode system is the delivery of sufficient current to protect part of the structure for the designed life of the system. There are three basic components of a marine galvanic anode cathodic protection system:

(1) Anode(s)
(2) Connecting System, including welding, bolting, cabling, and fasteners; and
(3) Protected Structure.

For requirements for the manufacture of galvanic anodes and detailed design of anode fastening, refer to *Cathodic Protection of Ships*.

3.5.2 Design Considerations

A detailed design of galvanic anode systems should, as a minimum, include the following:

(1) Acceptance criteria for the completed system
(2) Detailed drawings and specifications of anode alloys, sizes, and attachment
(3) Detailed calculations with specified design current density and anode resistance
(4) Detailed specification for inserts, attachment, and anode/structure continuity; and
(5) Detailed specification for installation, testing, commissioning, and operation.

3.5.3 Anode Material Properties

The current output of a galvanic anode depends on seawater resistivity and anode dimensions. The specific consumption rate of the anode is highly dependent on its environment. The anode life then can be calculated from the anodic material's consumption rate and its weight for a given current output. The dimensions, number, and distribution of anodes should be optimised in order to minimise the total weight of the galvanic anodes and to provide a protective electrical current greater or equal to the mean and maximum protection current demands required for the life of the anodes. The performance of a galvanic anode material (alloy) is dependent on its actual chemical composition and homogeneity, current density, and the environmental conditions to which it is exposed. In addition, anode surface morphology can affect the efficiency.

The electrochemical properties of anode material may include potential, current capacity, and anode consumption rate for the given environmental conditions. Where long-term performance data (of at least 12 months) is not available for a specific alloy and environmental combination, the properties of the anode material should be determined by appropriate tests, and caution should be exercised in selecting values for design purposes. Galvanic anode material performance is related to the chemical composition. Therefore, strict control of the alloy chemical composition of both the alloying elements and impurities is essential. The typical compositions of anode alloys should adhere to the recommendations found in this chapter, Sect. 3.3 of *Cathodic Protection of Marine Vessels* for anode materials of zinc alloys, aluminium alloys, and magnesium alloys, together with their close circuit potentials, practical current capability, and practical anode consumption rates.

3.5.4 Anode Arrangement for External Hulls

3.5.4.1 Installation Consideration

All anodes should be installed where the probability of disturbance to operations or mechanical damage is minimal. The anode assembly and its attachment should be designed to be highly resistant to mechanical damage. Generally, when few anodes are involved for high current outputs, the loss of an anode may significantly reduce the performance of the system. Anodes should not be located in areas:

- Where they can cause problems in the normal operation of the offshore structures
- That experience high stress or areas subject to high fatigue loads; and
- Where they could be damaged (by craft coming alongside, anchor chains or cables).

The number, dimensions, and location of anodes should be determined in order to be able to deliver the maximum protection current demand Imax and to achieve the cathodic protection criteria for the entire cathodic protection zone protected by that cathodic protection system. If suspended galvanic anodes are installed, the anode cables should be tested for strength, voltage drop, and electrical contact to the structure after installation. When separate suspension cables are used, care should be taken so that that anode lead wires are not in such tension as to damage the lead wires or connections. The anodes should be protected by covers with the visible words "DO NOT PAINT" when installed and during the coating process. If coatings are specified for anode supports or suspension cables, they should be visually inspected and repaired if damaged.

3.5.4.2   Anode Connection

The anode and anode core dimensions should be designed for the proposed fitting requirements. Anode cores should be fabricated from weldable structural steel and be compatible with the steel of the structure or structural elements to which they are attached. A low electrical resistance contact between the anode and the hull should be maintained throughout the operating life of the anode. Galvanic anodes can be attached to the structure either directly, by welding or bolting integral inserts to the structure (e.g., hull mounted anodes), or by a wire connection between the anode and structure. The welding on the structures should be performed in accordance with Class requirements. When the anode's steel inserts are attached to the hull by welding, stresses should be minimised at the weld location. The steel insert may be bolted to supports which have been welded to the structure. Attachment using studs "fired" into the structure is not permitted.

For anodes that are heavy (more than 230 kg [500 lb]) or subject to significant forces during installation and operation, a welded doubler plate or gusset plates with sufficient thickness is needed. In such case, an extra border of 20 mm (0.79 in) on all sides around the welding points of the anode should be provided. If installed as part of the anode fabrication, these plates may be subject to considerable damage during anode hauling and handling. Where anodes are to be attached to thin shell plates or sensitive materials or the hull plating of fuel oil or oil cargo tanks or internally coated areas, the use of doubler plates should be considered. If used, they should be continuously welded to the hull. Galvanic anodes may be attached directly to the hulls without shielding. However, a dielectric shield between the anodes and the hull metal can provide a better current distribution. For flush mounted anodes, a marine grade paint coating system [min. 100 μm (4 mils) DFT] should be specified for anode surfaces facing the protection object. If a wire is used, the wire connection to the anode should be completed by the manufacturer and multiple cables should be utilised to provide a degree of redundancy in case of cable damage. Cable attachment to the structure should provide both a mechanically secure and a low electrical resistance connection. Pin brazing, thermite welding, or "volcano" bolts or lock nuts with serrated washers as appropriate can be used.

3.5.4.3   Anode Arrangement

Anodes should be mounted on surfaces so as to avoid entrapping gas bubbles and disturbing water flow near intakes and discharge fittings. Anodes should be located so as not to disturb the flow of water past the propeller(s) or jet drive intake and nozzle(s). They should not be attached to areas likely to sustain regular mechanical damage (e.g., in way of the anchor). The anodes should not be attached to bottom plating or large unsupported panels and preferably not directly to the shell plating within the major longitudinal strength sections. The anodes should be uniformly distributed over the underwater surfaces to achieve good current distribution so as to protect the entire hull. For offshore structures in water with high conductivity (e.g., the Baltic) or requiring high protection current densities, the anode spacing should be smaller (such as 5 m (16.4 ft)). It should be reduced even further for the hull subjected to a large amount of mechanical damage (e.g., arctic waters).

3.5.4.4 Protection for Propellers, Shafts and Rudders

For the hull with a propulsion system installed, propellers, shafts, and struts should also be included in the cathodic corrosion protection of the outer shell. These parts should be connected conductively with the hull by means of slip rings on the propeller shafts and brushes. Refer to Chap. 2, Sect. 2.9 of *Cathodic Protection of Marine Vessels* for guidance on shaft grounding. If there is no central cathodic protection system, rudders should be cathodically protected by anodes, and propellers and shafts by zinc rings affixed to the propeller hubs or shafts. Sacrificial anodes on a shaft should be installed so as not to throw the shaft out of balance or restrict water flow to strut bearings. Collars applied to the propeller shaft are usually adequate to protect the propellers and shafts made from bronze and stainless steel. For Kort nozzles, a basic protection current density of 25 mA/m^2 (2.3 mA/ft^2) should be used for the total surface area. The anodes should be attached on the external surface at a spacing of $0.1r$ to $0.25r$ at the region of greatest diameter (r is the Kort nozzle's greatest radius). Internally, the anodes should be fixed to the strengthening struts.

The rudder should be included in the complete protection plan by cable or copper-band connections to the hull. If anodes are needed, anodes should be fixed on both sides either at the level of the propeller hub or as far as possible above and below the rudder blade. There are anodes specially designed for use on rudders, which are welded to the front edge of the rudder. Rudder heels should be given one anode on either side. The width of the anode should be smaller than the height of the rudder heel.

3.5.4.5 Anodes for Openings

The anodes should be arranged so that a shadow effect is avoided as much as possible. Openings in the outer shell, such as sea cheats, scoop openings, lateral thrust propellers, or similar, should be protected by externally placed anodes only up to a depth of one to two times the opening diameter. Where anodes are to be fitted for the protection of bow thruster units, they should be fitted as close as possible to the vulnerable areas, but with consideration to the water flow. A marine growth prevention system (MGPS) is usually installed in the sea chest. Cathodic protection current interference to MGPSs should be considered when anodes are fitted within sea chests.

3.5.4.6 Protection of Tubing and Piping

Galvanic or impressed current anodes are used to protect heat exchangers, condensers, and tubing components. The appropriate anode material is determined by the electrolyte. Zinc and aluminium are good for seawater and magnesium and aluminium are good for freshwater. Platinised titanium and mixed-metal oxide (MMO) can be used for the anode material in impressed current protection for seawater. Potential-regulating systems working independently of each other should be used for the inlet and outlet feeds of heat exchangers due to the different temperature behaviour. The protection current densities depend on the anode material and the fluid medium. The internal cathodic protection of pipes is only economically feasible for pipes with an internal diameter greater than 400 mm (15.75 in) due to the limit on range. Internal protection can be achieved in individual cases by inserting local platinised titanium wire anodes.

3.5.5 Cathodic Protection of Flooded Compartments and Ballast Tanks

3.5.5.1 General

Cathodic protection is required for freely flooded compartments and for closed compartments with free access to air. Those compartments include spud cans, ballast tanks, produced water tanks (elevated temperature) and magnetite-ballasted compartments. Closed and sealed flooded compartments do not normally require cathodic protection. The service condition of the tanks and compartments should be provided for design current density needed Cathodic protection in closed compartments without ventilation may cause development of hydrogen gas to an extent that an explosive gas mixture (i.e., hydrogen/oxygen) may eventually develop. The impressed current system may generate excessive hydrogen gas. However, the risk is moderate with aluminium- and zinc-based galvanic anodes. Only a galvanic anode cathodic protection system should be applied for flooded compartments and ballast tanks. Magnesium or their alloys are not acceptable anodes in seawater tanks, except in freshwater tanks that are not adjacent to cargo tanks.

The compartments' and tanks' size, shape, and areas to be protected should be presented accurately in the cathodic protection design. Type of steel and areas coated/uncoated should be considered. For ballast tanks, the ballasting routines, including the ballasting percentage of the total time, the probable duration of ballasted periods, and level of ballast water should be considered. A cathodic protection system is not effective when the tank/compartment is empty. It takes some time (½ day or more) to obtain re-polarisation of submerged steel surfaces after filling with seawater. Unless the tank/compartment is completely filled with seawater, the ullage space or under deck area on top of tanks is not protected by the anodes.

3.5.5.2   Requirements for Coating

A tank's and compartment's coating qualification and installation should meet requirements from the coating producer's recommendation and/or regulatory/class requirements, such as IMO Resolution MSC.215(82), "Performance Standard for Protective Coatings for Dedicated Seawater Ballast Tanks in All Types Of Ships and Double-Side Skin Spaces of Bulk Carriers". The coating system should resist cathodic disbonding and pass the test criteria when it is required.

3.5.5.3   Protective Current Density

The recommended design current densities for freely flooded compartments, closed compartments with free access to air, and seawater ballast tanks are given in this chapter, Table 3.2.

Generally, the mean current density is about half of the initial current density (i.e., $I_{cm} = \frac{1}{2}$ of I_{ci}). The final current density is about 2/3 of the initial current density (i.e., $I_{cf} = 2/3$ of I_{ci}). The basic cathodic protection design for the ballast tanks is based on the combination of cathodic protection and a coating. The cathodic protection current demand can be influenced by temperature, sloshing quality of the seawater, complex tank/compartment geometry, and coating breakdown. The coating breakdown factor (f_c) can be estimated by following equations when there is no other data available:

The mean coating breakdown factor, $f_{cm} = 0.01 + (0.0048\,to\,0.006)t$
The final coating breakdown factor, $f_{cf} = 0.01 + (0.012\,to\,0.0145)t$.

The cathodic protection current demand calculations should be based on utilising one of the two design alternatives for design life:

- Cathodic protection current demand design calculation should be limited to $I_{total\;mean} \geq I_{cm}$; and

Table 3.2   Recommended initial current densities for bare steel surface exposed to seawater with a range of water temperatures

Steel surface water temperature, °C (°F)	Initial I_{ci}, mA/m^2 (mA/ft^2) for bare steel
>20 (68)	120 (11.1)
>12–20 (54–68)	140 (13.0)
>7–12 (44–54)	170 (15.8)
≤7 (44)	200 (18.6)

- To minimise the owner's inspection, maintenance, and repair (IMR) and avoid any retrofitting of anodes in seawater ballast tanks/compartments, cathodic protection current demand design calculation should include the calculations of both the mean current demand (I_{cm}) and the final current demand (I_{cf}) for meeting $I_{total\,final} \geq I_{cf}$.

The owner should decide if the cathodic protection design for the seawater ballast tanks/compartments is to be based on the alternative.

3.5.5.4 Anode Weight

Anode materials should be selected in accordance with this chapter of *Cathodic Protection of Ships*. The required anode weight per zone can be calculated by:

$$M_{zone} = I_{zone} \times t_s \times \frac{f_B}{Q_g} \text{kg[lb]}$$

where

I_{zone}	total protective current $I_{zone} = A_{zone} \times i_{zone}(\text{A})$
M_{zone}	area of a cathodic protection zone, in m^2 (ft^2). The maximum surface area covered by the electrolytic solution is used for the calculation
i_{zone}	necessary protective current density for a cathodic protection zone, in A/m^2 (A/ft^2)
T_s	protective period. The protective duration should be set to 5 years (43,800 h) or defined by agreement with the designer or owner
f_B	loading (utilisation) factor, which depends on the period in which the surface is covered with the electrolytic solution. In the case of constant loading (filled tanks/cells), the factor is to be set to 1
Q_g	electrochemical efficiency of the anode alloy, in A h/kg [A h/lb].

3.5.5.5 Anode Arrangement

Anodes should be arranged so that a shadow effect is avoided, including areas with complex structures. The arrangement should also consider the prevention of fire and explosion. Because of uncertain or varying water levels, anodes should be primarily placed in the lower part of a tank/compartment, where the surface is most likely to remain wet. It should be noted that several smaller anodes provide a superior current distribution than one large anode of the same total weight. It may be necessary to increase the number of anodes for the internal spaces for the following reasons:

- The effective zone of the anodes for low water levels
- A shadow effect caused by complex internal structures
- Galvanic corrosion from use of noble materials, which should be compensated locally; and

- It may also be necessary to provide extra anodes in addition to the total anode weight calculated in order to achieve the required protective current.

Anodes should not be located under tank hatches or butterworth openings unless protected by the adjacent structure. For seawater applications, aluminium anodes should be used. Aluminium anodes should be located in such a way that they are protected from falling objects. Aluminium anodes can only be used for liquid cargo with flash points below 60 °C [140°F] provided aluminium anodes are located so a kinetic energy of not more than 275 J [203 ft-lb] is developed in the event of their loosening and becoming detached. This limitation does not apply for ballast water tanks that are not adjacent to cargo tanks. There is no height restriction for zinc anodes.

3.5.5.6 Anode Attachment

Anodes should be attached to stiffeners or aligned in way of stiffeners on bulkhead plating, but they should not be attached to the shell. The two ends of the anode should not be attached to separate members which are capable of relative movement. Where cores are welded to local support members or primary support members, they should be kept clear of end supports, toes of brackets, and similar stress raisers. Where they are welded to asymmetrical members, the welding should be at least 25 mm (1 in) away from the edge of the web. In the case of stiffeners or girders with symmetrical face plates, the connection may be made to the web or to the centreline of the face plate, but well clear of the free edges. Generally, anodes should not be fitted to a face plate of a higher strength steel. The steel insert of galvanic anodes can be bolted to separate supports (brackets) connected to the stiffeners by continuous welding. Anodes can also be attached to flanged stiffeners by the use of bolted clamps. In such cases, the clamping bolts should be fitted with additional locking nuts. Electrical continuity checks should be performed, and the resistance should be such that the IR drop across any bolted connection is less than 10% of the design driving voltage between anode and steel structure, and in no case higher than 0.1 Ω. Tanks/flooded compartments in which anodes are installed are to have sufficient holes for the circulation of air to prevent gas from collecting in pockets.

3.5.5.7 In-Service Inspection and Retrofitting

The ballast tanks and other tanks/flooded compartments with seawater are subject to inspection in 10 years of service life or longer. Depending on the owner's inspection, maintenance, and repair (IMR) strategy, these spaces may be available for easy and cost-effective retrofitting of sacrificial anodes. There is no requirement for a permanent monitoring system of a sacrificial anode system. Manual monitoring of the effectiveness of the system may be done by visual inspection of the anodes and/or by measurement of the protection potential. Measurement of protection potentials should be performed using a portable reference electrode. It is recommended that a plan for retrofit of anodes be developed when the measured potential is more positive than −900 mV (Ag/AgCl/

Seawater), and anode replacement is needed immediately when measured potential is more positive than −800 mV (Ag/AgCl/Seawater).

3.5.5.8 Lay-Up Period

During a long lay-up, ballast tanks and other tanks/compartments with seawater should be maintained either full or empty. The tanks are to be kept empty and dry, or full, with a sacrificial anode cathodic protection system installed.

3.6   Impressed Current Cathodic Protection

3.6.1   General

Unlike galvanic anode systems where the natural potential difference between the anodes and the steel surface provides the driving force for current, an impressed current cathodic protection (ICCP) system is supplied with power from an external source. The electrical current output delivered by the DC power source (normally a transformer rectifier) is controlled during the lifetime of the cathodic protection system in order to obtain and maintain an adequate electrochemical potential level for protection over the whole hull to be protected. Since ICCP systems predominate in modern ships because of their automatic adjustment for changes in ship speeds and electrolyte resistivity (such as when a ship enters brackish or fresh water), FPSOs converted from oil tankers tend to use the existing ICCP system installed. Present experience in offshore floating structures indicates that for a 15-year or longer service the sacrificial anode cathodic protection (SACP) system are considered to be the most effective alternative for the underwater hulls of the offshore floating structures. Combination systems may be required and a SACP system should be used for protection prior to the commissioning of an ICCP system and/or additional sacrificial anodes to dedicated items (e.g., for sea chests) may be required. ICCP systems are not to be used in ballast tanks or other tanks due to development of hydrogen, which can cause an explosion hazard.

3.6.2   ICCP System Components

The ICCP system primarily includes the following components:

- Anodes
- Power source
- Connecting system, including cables, connections, bonding
- Monitoring and control systems
- Dielectric shields

- Reference electrodes
- Stuffing tubes and cofferdams; and
- Protected structure.

Chapter 4, Sect. 4.2 of *Cathodic Protection of Marine Vessels* provides details of the ICCP components.

3.6.2.1 Transformer Rectifier Power Source, Monitoring, and Control Systems

Impressed current cathodic protection systems for offshore structures usually include one or more transformer rectifier(s) along with multiple anodes and reference electrodes. The DC power source should be able to deliver sufficient current to provide the adequate protection potential of the cathode. The current capacity (It) of the cathodic protection system should be designed to be able to provide at least 25% more than the calculated current demand, I_{max} (i.e., $I_t \geq 1.25I_{max}$). The transformer rectifier output voltage should take into account the voltage drops from the resistance of the electric circuit and the recommended operating voltage of the anodes. Transformer rectifiers with automatic potential control are generally used to meet the current demand necessary to maintain structure polarisation because the environmental conditions and the coating conditions are frequently varied. The transformer rectifier should be equipped with monitoring and control equipment. For an automatically controlled DC power source the potential control should be able to deliver a current when the control unit reads a potential more positive than the set potential limit. Similarly, the DC power source should not deliver current when the control unit detects a potential more negative than the set potential limit. There should be the ability to limit the current output from each anode to a preset value.

3.6.2.2 Anodes

Unlike galvanic anodes, impressed current anodes are designed to be resistant to corrosion. Desirable properties include low resistance to current flow, physical toughness, low rate of consumption, and low cost of production. Platinum is an ideal candidate for impressed current anodes because it has an exceptionally low consumption rate, and thus longer service life. It is cost-effective to use platinum-coated titanium, tantalum, or niobium rather than solid platinum for the anodes. The mixed metal oxides (MMO) are also usually used as the inert anodes. The number, size and location of the anodes should be determined to be adequate for delivering the required current from the DC current source. The choice of anodes depends on the expected severity of operating conditions together with cost and durability considerations. Generally, the current outputs for the inert anodes vary from about 400 to 3000 A/m^2 (37–278 A/ft^2). The current output is inversely proportional to the grounding resistance for a given voltage. This is proportional to the conductivity of the water and can vary by a factor of one hundred. Thus, the anode voltages should be correspondingly raised in poorly conducting waters to achieve the required protection current densities. In high-resistance waters, under protection can occur as a result of voltage limitation of the anode used. MMO anodes have no such voltage limitation as titanium coated anodes. The protection should be designed for the particular type of water. Precautions should be taken to avoid a short circuit of the anode to the structure and water leakage at the anode penetration. A hull penetration with a cofferdam and a sealing arrangement is normally used for the mounting of the anode.

3.6.2.3 Reference Electrode

Reference electrodes should be permanently installed at locations determined by calculation or experience so that the potential of the hull is maintained within the set limits. Reference electrodes are essential for the control of an ICCP system. Reference electrodes are used to measure the potential of the structure (steel) for controlling the electric current delivered to the cathodic protection system. Normally, two reference cells are installed approximately halfway between the anodes powered by the same controller. One acts as a primary control, while the other serves as an auxiliary to verify operation of the primary cell. The auxiliary reference cell is important for verifying system operation. It can provide primary service if the first cell fails. Reference electrodes are to be replaceable. They can be designed for electrode change-out while afloat. The reference electrodes are available in zinc and silver/silver chloride materials. In arctic waters, the reference electrodes attached to the hull and their arrangements should be recessed into the hull to protect from heavy mechanical wear caused by ice.

3.6.2.4 Wiring and Connections

In ICCP systems, all wiring and connections should be totally isolated from the electrolyte. Approved cathodic protection dielectrically insulated cable can be used. Unlike a galvanic anode system, where exposed wire and connections are protected by the anode, any exposed metal in the electrolyte is part of the anode in the ICCP system and will corrode rapidly. The cross-section of the cables should consider possible voltage drops for the length of the cable. The specified maximum current rating of the cable should not be exceeded. Dedicated cables should be used for potential measurements. Such cables should be suitably screened/sheathed to avoid interference.

3.6.2.5 Stuffing Tube(S)

Stuffing tubes are required for impressed current anodes and reference electrodes. If a stuffing tube penetrates a fluid-filled compartment, such as a fuel tank, bilge, ballast tank, or freshwater tank, the stuffing tube and the electrical cable leading to it should be enclosed in a watertight cofferdam. Requirements on stuffing tubes are addressed in MIL-S-23920, Stuffing Tubes for use with Circular Anodes and Reference Electrodes (Corrosion Preventive). Typically, the stuffing tubes are supplied with the reference electrodes and anodes as assemblies.

3.6.2.6 Cofferdams

Hull anodes, reference electrodes, and other components that penetrate the hull below the waterline should be designed, constructed, and installed so that the mechanical integrity and watertightness of the hull should be maintained. Cofferdams should be used to facilitate the entry of the cables into the interior of the hull. The construction materials used for cofferdams should be metallurgically compatible with the hull. The steel plate thickness of these cofferdam boxes should correspond to that of the hull. The manufacturing and fitting of cofferdams is to be in accordance with relevant international, national, or classification society requirements.

3.6.2.7 Dielectric Shield

The current at which the anodes are required to operate will result in unduly negative potentials immediately adjacent to the anode. These potentials may result in disruption to conventional coatings and can be detrimental to high-strength steels by causing hydrogen embrittlement. Consequently, a heavy-duty dielectric shield is needed in the area of the hull immediately behind and adjacent to the active anode element to prevent shorting of current to the hull and to aid in wider and more even current distribution. The dielectric shield materials selected should be resistant to cathodic disbonding, corrosive chemicals produced at the anodes, as well as significant deterioration or aging. Any damage to the shield will result in unacceptably high negative potentials on the exposed hull.

High-performance dielectric shields (coatings) are often applied directly to the hull surface prepared in accordance with the coating manufacturers' recommendations. Prefabricated shields formed from glass (or fibre) reinforced polyester/epoxy or thermoset plastics, either as thin sheet materials pre-bonded to a steel doubler plate or as a direct attachment to the hull, can be used. Integral anodes and dielectric shields are available for large flat areas to avoid the need for periodic maintenance of the coatings. Typically, the standard BS 7361 is typically used to help determine the shield dimensions. The size of the dielectric shield will be determined by the shape of the anode, cathodic protection electrical potentials, the maximum current output, and the resistivity of the seawater.

3.6.3   ICCP Design Considerations

The design calculations and specifications should include the following detailed information:

- Design basis
- General arrangement of the equipment
- Specification of equipment, such as the power source, equipment size, monitoring and control systems, anodes, connection cables, reference electrodes, and connection and protection devices

- Specification for equipment installation; and
- Specification for monitoring system.

When applicable, each cathodic protection zone (refer to Chap. 2, Sect. 2.6.2) should be protected by a dedicated system. Specific areas may require a multi-zone control system in order to adapt and optimise the electrical current distribution to the cathodic protection demand. The anode distribution should provide an even current distribution around the hull during the cathodic protection design life. The impressed current system design and installation, including anodes, reference electrodes, rectifiers, cabling, hull penetrations, cofferdams, monitoring units and anode shields should be supplied by recognised providers. During the fitting and installation period that electrical current supply may not be available on the structure and the impressed current system cannot be activated, the hull should be protected by a sacrificial system temporarily. The temporary sacrificial system can use the anodes hanging along the hull side.

3.6.4 ICCP Installation Considerations

The impressed current cathodic protection system installation should be in accordance with the following:

(1) All electrical connections to the AC or DC electrical system should comply with applicable codes and operator's specifications. Nameplate and actual rating of the DC power source should comply with construction specifications. An external disconnect switch in the AC wiring to the rectifier should be provided
(2) Impressed current anodes should be inspected for compliance with specifications for anode material, size, and length of lead wire and to confirm that the anode cap (if specified) is secure. Lead wires should be carefully inspected to detect possible insulation defects. Defects in the lead wires and anode caps are to be properly repaired, or the anode should be rejected. Insulation damage from construction and installation can arise (e.g., due to welding) and should be immediately repaired. Careful supervision of this phase is essential for proper long-term performance of the cathodic protection system
(3) The anode and electrode surfaces should be protected by covers with the visible words "DO NOT PAINT" when installed and during the coating process
(4) Conductor cable connections to the rectifier from the anode(s) and the structure should be mechanically secure and electrically conductive. Before the power source is energised, verification is to be made that the negative (–) conductor is connected to the structure to be protected, that the positive (+) conductor is connected to the anode(s), and that the system is free of short circuits (i.e., the anode is NOT shorted to the structure)

(5) System controllers should be installed at least 1 m (3 ft) from compasses and other magnetically sensitive equipment. Associated wiring should be either twisted pair type or shielded when in proximity to such equipment

(6) A rectifier or other power source should be installed out of the way of operational traffic and remote from areas of extreme heat or likely contamination by mud, dust, water spray, and other contaminants. It should be located outside of electrically classified areas. Where two or more rectifiers are installed, the rectifiers should be spaced to allow proper flow of cooling air

(7) Openings in the outer shell, such as sea chests, overboard discharges, stabiliser boxes, thrusters, scoops, parts not conductively linked, and shaft penetrations, should be protected additionally with sacrificial anodes

(8) When installing a suspended anode for which a separate suspension cable is required, care should be taken that the lead wire is not in such tension as to damage the anode lead wire or connections; and

(9) Operating personnel should be trained to maintain continued energisation of the system.

3.6.5 Cathodic Protection in Service Monitoring

For carbon steels and austenitic stainless steels as examples, when measured polarised potential is more negative than -1000 mV (Ag/AgCl/Seawater), the "set" values of the ICCP system is required to be checked. When the measured polarised potential is more positive than -900 mV (Ag/AgCl/Seawater), ICCP system anode current output should be increased without exceeding the polarised potential negative limit -1100 mV. For other materials, measured polarised potential should be controlled within the allowed range of potential. It is important to note that using a hybrid cathodic protection system (utilising both SACP and ICCP) for turrets, sea chests, moorings, or risers, the SACP system potential may deliver a potential high enough to cause the ICCP system to shut down.

Commissioning, Operation, and Maintenance

4

4.1 General

The objectives of the commissioning, operation, and maintenance of the cathodic protection system are that the:

(1) Cathodic protection system functions continuously in accordance with the requirements of the design and installation; and
(2) Hull, internal surfaces (ballast tanks), and appendages remain satisfactorily protected from corrosion over the life of the system.

4.2 Potential Measurement

The protection criteria and effectiveness of cathodic protection systems should be confirmed by direct measurement of the structure potential. The potential measurement techniques and procedure can be found in Chap. 2, Sect. 2.4.2. The cathodic protection potential measurements should be conducted by the qualified cathodic protection personnel as stipulated in Chap. 1, Sect. 1.5. Measured potential must be referenced to the specific reference electrode suitable for the electrolyte.

4.3 Commissioning: Galvanic Systems

For galvanic systems, the hull/seawater potentials should be measured within one (1) month of the installation. The potential survey should be done by using a portable reference electrode to supplement any permanently installed reference electrode. The survey

© The Author(s), under exclusive license to Springer Nature Switzerland AG 2025

A. A. Olsen, *Cathodic Protection of Offshore Structures*, Synthesis Lectures on Ocean Systems Engineering, https://doi.org/10.1007/978-3-031-77581-9_4

should check that the criteria selected at the design stage in Chap. 2 are met at representative locations. Ballast tank system potential measurements should be made in conjunction with the ballasting program.

4.4   Commissioning: Impressed Current Systems

4.4.1   Visual Inspection

In the drydock, the electrical isolation of the anodes from the hull should be confirmed by electrical resistance measurement. All cable circuits should be checked for continuity and insulation. The polarity of the DC output should be confirmed. The cathodic protection system and all its components should be subject to a complete visual inspection within the drydock to confirm that all components and cables are installed properly, are labelled where appropriate, and protected from any possible damage. The anodes and reference electrodes should be visually inspected to confirm that they are not coated, and the installation is in accordance with the design documentation. The dielectric shield around the anodes should be visually inspected to confirm that the installation is done appropriately as detailed in the design documentation. The shield should be tested for film thickness and for absence of defects ("holidays") both in accordance with the specification and the dielectric shield manufacturer's recommendations. All inspection data should be recorded and maintained.

4.4.2   Pre-energising Measurements

Before the cathodic protection system is switched on when the vessel is floated out from the drydock, pre-energising measurements should be made as soon as possible, including:

(1) The potentials of the hull and appendages to seawater with respect to all permanently installed reference electrodes
(2) The potentials of the hull and appendage to seawater with respect to portable reference electrodes; and
(3) Any electronic data logging and/or data transmitting facility installation as part of the performance monitoring system.

4.4.3   Initial Energising

The cathodic protection system is to be energised in accordance with the design requirements for initial polarisation. Measurements should be made including:

(1) The potentials of the hull and appendages to seawater with respect to all permanently installed reference electrodes

(2) The output voltage and current of all transformer rectifiers and the current of all individual anodes; and

(3) If any steel/seawater potential value shifts in a positive direction, it should be investigated to determine any requirements for additional testing and/or remedial work.

4.4.4   Performance Assessment

Within one (1) month of energising, a survey of potential measurement should be undertaken using a portable reference electrode to supplement the permanent monitoring provisions installed to confirm that the design criteria are met at representative locations. A repeat survey as above should be undertaken one (1) month before the end of the Defect Liability Period (or guarantee) for the vessel and its cathodic protection system or within twelve (12) months of drydocking and anode replacement.

4.5   Operation and Maintenance

4.5.1   General

The operation and maintenance testing intervals and procedures should comply with the operation and maintenance manual or as subsequently modified based upon performance of the system for continuous, effective, and efficient operation of cathodic protection systems.

4.5.2   Galvanic Anode Systems

For galvanic anode systems, a periodic performance assessment should be undertaken by taking potential measurements at identified locations around the hull. Following the commissioning testing denoted in this chapter, Sect. 4.3, further testing should be performed, typically between nine and twelve months and then at intervals of two to five years, subject to proven design life and planned dry docking intervals. In addition, depending on the type of vessel, locations of anodes and planned dry docking intervals, a visual inspection of the anodes may be undertaken by underwater survey inspection. The survey should assess the consumption of the anodes and check physical damage to anodes. Any damaged, consumed, or missing anodes should lead to potential surveys and estimation of remaining life of anodes for the protection level of the hull or appendages. If necessary, these anodes should be replaced or retrofitted.

4.5.3 Impressed Current Cathodic Protection Systems

For impressed current systems, normal operation includes confirmation that:

- The system is switched on and all systems are functioning
- A record of the system operation is in place with any down time recorded; and
- All anode current outputs are similar to those found during the previous assessment.

Overall monitoring and inspection procedures should include:

- Measurement and recording of transformer rectifier output total current and voltage (regularly)
- Measurement and recording of hull steel/seawater potential with respect to permanently installed reference electrodes (regularly)
- Measurement and recording of anode current outputs (regularly)
- Measurement of parameters from any other sensors installed as part of the performance monitoring system (as appropriate)
- Calibration of permanent reference electrodes (i.e., annually). A portable reference electrode should be used for their calibration; this should be placed as close as possible to the permanent measurement electrode and the cathodic protection current should be switched off during the calibration procedure
- A detailed representative survey of the entire structure using portable reference electrodes annually or after any major repair/refurbishment/maintenance of the cathodic protection system; and
- The measurement of potential difference between anodes and the hull to verify the metal-to-metal isolation of anode to hull.

Impressed current systems can pose a risk to divers and are normally switched off during diving operations in their vicinity. If this is impracticable, divers are to be informed so that the necessary actions can be taken with respect to their safety.

4.6 Planned Drydocking or Underwater Survey Period

Coatings should be visually examined to determine if the coating deterioration is within the value assumed in the cathodic protection design, and if there is any evidence of coating damage caused by the cathodic protection. When out-of-water drydocking survey is needed, impressed current systems should be checked with insulating resistance of anodes and electrodes to the hull. Measurements should be carried out after having cleaned the periphery of anodes and electrodes to avoid an electrical continuity due to salt deposits. Insulating resistance should be more than 1 MΩ. Values below 1 MΩ may be acceptable for performance but should be investigated as they indicate possible deterioration. When drydocking or underwater survey (UWILD) is performed, galvanic anodes should be inspected and replaced/retrofitted if their consumption rate is not adequate for the design life.

4.7   Fitting Out and Lay-Up

4.7.1   General

Precautions should be considered at the design stage to provide adequate protection of underwater surfaces, including appendages, propellers, and shafts, during shipyard's fitting-out time for newbuilds, refitting time, repair time, or lay-up periods. The choice between suspended galvanic anodes and an impressed current system powered from a shore supply should be determined by the availability of supply and whether occupation of a berth is anticipated.

4.7.2   Fitting-Out Period

The fitting-out period can last up to several months depending on the type of vessel. Conditions in fitting out berths are often severely corrosive. It is important that cathodic protection is applied during this period to prevent corrosion. If a galvanic anodes cathodic protection system is to be used in service, it should be fitted before launching. If an impressed current cathodic protection (ICCP) system is to be used in service, temporary galvanic anodes should be installed and activated before the ICCP system is installed or activated. The temporary galvanic anodes should be bonded electrically to the hull and may be suspended at regular intervals around the hull. These anodes should be sufficient to provide full polarisation and hull potential should be maintained and monitored at a satisfactory level.

4.7.3   Lay-Up Period

One of the basic criteria in designing lay-up procedures is the preservation and maintenance of the hull and machinery with appropriate corrosion protection. External coating systems should be in good condition prior to lay-up. The underwater hull area should be adequately protected with suspended sacrificial anodes or the impressed-current cathodic protection system if a continuous power source is available for use. Hull potential should be maintained and monitored at a satisfactory level. At permanent moorings, galvanic or impressed current anodes may be laid on the seabed, provided the clearance between the anodes and the keel at low water is sufficient to avoid paint damage. Hulls that are laid up and static for extended periods may be subjected to marine growth and microbially influenced corrosion (MIC). In these circumstances, the protection criteria for anaerobic conditions may be used. The initial setting of the correct current requires two or three cathodic protection potential surveys at intervals of a few days. Thereafter, the surveys may be at intervals of several months, provided the water conditions remain stable and the operation of the cathodic protection system is stable. The proximity of other vessels or structures and the need for interaction testing should be considered.

4.7.4 Stray Current

During the fitting out and lay-up periods, some onboard hot work repairs may be requested. In that case, the stray current caused from welding process may cause unpredictable damage to the underwater hull and shaft. It should be noted that underwater hull/structures/shaft may experience serious corrosion induced by the stray electrical currents from improper weld lead hookups and/or from using onshore power.

Retrofit of a Cathodic Protection System

5

5.1   General

As noted in Chap. 3, Sect. 3.6.5, a plan for retrofit of sacrificial anodes is needed when the measured polarised potential is more positive than −800 mV (Ag/AgCl/Seawater) for carbon steels as an example. Retrofitting of sacrificial anodes for offshore structures are generally very costly and sometimes impractical. It is therefore normal practice to have at least the same design life as for the structure with minimal requirements for maintenance. However, in certain circumstances planned retrofitting of sacrificial anodes may be an economically viable alternative to the initial installation of an exceptionally large amount of anodes. This alternative retrofitting should be planned during the initial design and fabrication. Depending on the owner's inspection, maintenance, and repair (IMR) strategy, the ballast tanks and other tanks may be available for easy and cost-effective retrofitting of sacrificial anodes. When the economics of a cathodic protection retrofit are considered, the installation cost always dominates, which includes daily costs from diving/ROV, the support staff and support vessel. The economics and apparent simplicity of design and installation offered by impressed current systems have always attracted cathodic protection designers. Current requirement and water depth may also determine if impressed current cathodic protection (ICCP) system is to be used. As a general rule of thumb, impressed current systems are to be considered attractive when current requirements exceed 400 A and/or when water depth exceeds 70 m (210 ft). The following should be considered early in the decision process:

- Safety concerns for divers, fishers, and wildlife
- Supply of sufficient spare current capacity to power the ICCP system
- Sufficient space to install the transformer rectifier(s)
- Cable routing options for the structure ground (−) and anodes (+)

© The Author(s), under exclusive license to Springer Nature Switzerland AG 2025
A. A. Olsen, *Cathodic Protection of Offshore Structures*, Synthesis Lectures on Ocean Systems Engineering, https://doi.org/10.1007/978-3-031-77581-9_5

- Water depth for applicable retrofit system
- Prequalification of the retrofit system to be used by evaluating the performance data, and a written warranty
- Risk of interference posed by pipelines on and around the structure
- Location of ICCP anodes so as to optimise performance (current distribution and cable runs) and avoid interference or overprotection
- The structural nature for ICCP. Extremely complicated structures, or the use of extremely high strength steels may prevent the use of ICCP, or make installation unpractical
- Availability of cathodic protection personnel to operate and maintain the system
- Monitoring and inspection of the system
- A comparison of a galvanic retrofit design and ICCP retrofit design should be made. If the installed cost of the ICCP system is not significantly less, the galvanic retrofit system should be chosen; and
- If structure has galvanic anodes, attempts should be made to schedule the retrofit to make use of the remaining current capacity.

5.2 Cathodic Protection Retrofit Systems

5.2.1 General

Offshore harsh conditions make traditional cathodic protection technology for retrofitting difficult and costly. Special retrofitting technologies are used for specific applications. For any vessel, whether new-build or conversion, a sacrificial anode system is recommended if the expected life is greater than 10 years. This recommendation is supported by both economic and reliability studies. For depleted systems requiring offshore retrofit the recommendation is usually to deploy impressed current due to installation costs. For FPSOs, a deep suspended system would almost certainly be the most cost-effective if 360° rotation around a turret is allowed, or a seabed-deployed system is sometimes the best option. As an example, an FSU vessel, which is spread moored, was retrofitted with an impressed current retrofit system. The original sacrificial anodes were virtually depleted, and offshore replacement of the depleted system would be cost prohibitive due to the extended amount of diving activity required. After study of several types of retrofit systems, the remote buoyant anode, located on the seabed with feed cables was selected. The advantages of this system are:

- Only two anodes on two sleds (at 400 A each) are needed, which provide uniform current distribution over all the protected areas of the vessel and its appurtenances

- If the anodes are sufficiently removed from surrounding structures such as pipelines, subsea support structures, and production equipment, there is no risk of stray current interference since there are no locally high voltage gradients present; and
- Installation can be accomplished in one or two days using ROV support only.

5.2.2 Hanging Impressed-Current Anode Systems

Hanging Impressed Current Anode Systems have been successfully applied to offshore floating storage installation vessels originally using sacrificial anodes for the hull. In this system the anode is freely suspended from locations above water either from the feed cable or a strain member. The anode is often weighted and contained within a dielectric frame. The hanging anode can be retracted from a porch during severe weather. Refer to Fig. 5.1. This can be a viable approach for shallow water structures with a relatively short life expectancy (<5 years) and a moderate to high current demand. Anode failure is expected on a regular basis, but with exceptionally low replacement cost. To provide even current distribution to the hull, the anode of the cathodic protection should be hung at least a few metres (about 10 ft) away from the hull. This system could be installed without interference to offshore production and operations.

5.2.3 Gravity Anode Sled

Most gravity sleds are galvanic anodes. Refer to Fig. 5.2. The anode sleds are designed to sit on the sea floor at some distance from the structure and connected electrically to the protected structure. Most failures of these systems stemmed from damage to the seabed cable or cable connection failure of the anode.

5.2.4 Anode Pod and Anode Mat

The anode pod is an aluminium anode system arranged in stable, self-contained "pods" which are ideal for replacing depleted anodes on mature assets. The pods are lowered to the sea floor and connected electrically to the protected asset via the clamp system. The anode mats function in the same way but are more stable in shallow water and serve to stabilise subsea structures. For more information refer to Fig. 5.3.

Fig. 5.1 Hanging anode
system at retracted position for
a FPSO

(Photo courtesy of Deepwater Corrosion Services)

5.2.5   Anode Link

The anode link is a cost-effective cathodic protection retrofit system consisting of three
to fifteen anodes cast directly onto a heavy-duty wire rope. Refer to Fig. 5.4. The anode
link attaches electrically and mechanically above the waterline allowing the string to hang
in the seawater with at least two anodes situated in the mud. The installation is almost
always performed from a small vessel without using divers. This solution is typically
suitable for any structure in less than 30 m (90 ft) of water.

(Photo courtesy of Deepwater Corrosion Services)

Fig. 5.2   Anode sled

5.2.6   Buoyant Sled

A buoyant sled is a high-capacity impressed-current cathodic protection system. Refer to Fig. 5.5. It utilises impressed-current titanium-anode rods housed in buoyant floats. Similar to gravity sleds but with the anode elements held up in the seawater by means of buoys, the sled is normally integral with the anodes. The advantage of the buoyant sled is that the critical elements can move freely if hit by falling debris, and the overall sled structure is much lighter. Each buoyant anode sled can be rated from 150 to 500 A to maintain appropriate levels of cathodic protection. The compact size and remote location of the buoyant sled on the seabed simplifies the installation procedure and increases the cost-effectiveness of the retrofit.

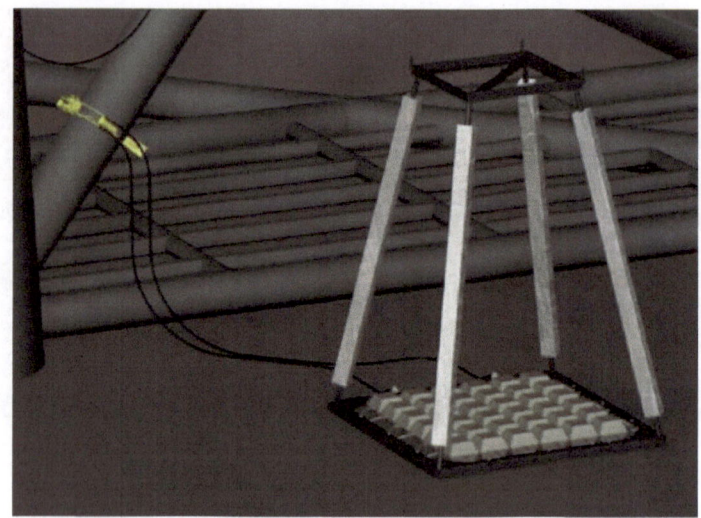

(Photo courtesy of Deepwater Corrosion Services)

(Photo courtesy of Deepwater Corrosion Services)

Fig. 5.3   Anode Pod and Mat

Fig. 5.4   Anode link

(Photo courtesy of Deepwater Corrosion Services)

Fig. 5.5   Buoyant sled

(Photo courtesy of Deepwater Corrosion Services)

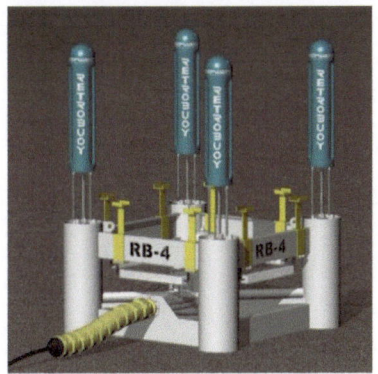

(Photo courtesy of Deepwater Corrosion Services)

Documentation

6

6.1 General

The design, installation, energisation, commissioning, long-term operation, and maintenance of the cathodic protection system should be fully and permanently recorded. The initial installation documentation should reflect the latest revision of design specification including the equipment locations, and waterline. Commissioning data should include results of the cathodic protection surveys conducted after energising each cathodic protection zone, including structure potential measurements and any interaction testing with respect to adjacent structures.

6.2 Galvanic Anode Systems

The following data should be maintained for reference and should be updated if, and when, changes are made to the system:

(1) Design criteria including the design life, environment characteristics (i.e., water resistivity, etc.), protection criteria, current density requirements, anode current output values at different periods and working conditions, and the documented amp-hour capacity of anodes
(2) The number of anodes, their dimensions, weight, specification, alloy composition, documented amp-hour capacity, open circuit potential and other characteristics, as well as the manufacturer/supplier references and documentation

A. A. Olsen, *Cathodic Protection of Offshore Structures*, Synthesis Lectures on Ocean Systems Engineering, https://doi.org/10.1007/978-3-031-77581-9_6

(3) The anode locations marked on a specific drawing of the structure, the method of attachment, date of outfitting afloat, date of installation, and date of actual setting on offshore location, with all discrepancies noted. These data should be updated during the life of the vessel/structure

(4) the location, description, and specification of any current or potential control or monitoring devices, including reference electrodes, measuring equipment, and connecting cables

(5) The commissioning results, including potential survey data from both fixed reference electrodes (if any) and portable reference electrodes for representative survey of the entire structure

(6) The cathodic protection survey results of periodic maintenance, including current (if possible) and protection potential measurements, equipment, and the measuring technique to follow the changes of the protection potential status of the structure; and

(7) An Operation and Maintenance Manual which should detail the as-built system, inspection and testing procedures, and inspection and testing intervals.

6.3   Impressed Current Systems

The following data should be recorded and maintained for reference and be updated if, and, when changes are made to the system:

(1) The design criteria including the design life, environment characteristics (such as water temperature range, salinity range and resistivity), protection criteria, current density requirements, design values of the anode output current and associated power supply output voltage at maximum current and anticipated operating currents at minimum and maximum extent of coating breakdown

(2) The number and specification of anodes, including their dimensions, anode element composition, connection details, anode current densities and voltages, maximum/average/minimum anode life, as well as the manufacturer/supplier information

(3) The attachment details of anodes and reference electrodes and the specification of the connecting cables and their through hull arrangements

(4) The locations of each anode and reference electrode as confirmed during construction, with all discrepancies noted, and the date of installation. This data should be updated during the life of the structure

(5) The specification of any dielectric shield used, including the location, dimensions, surface preparation, material, dry film thickness, and inspection data recorded during the installation of all dielectric shields

(6) The location, detailed specification, drawings, circuit diagrams, and output characteristics of each DC power source (such as transformer rectifiers) and their factory test reports

(7) The location, description, and specification of any performance monitoring and control system, electrical protection devices (fuses, circuit breakers, etc.), measuring equipment, and connecting cables

(8) The commissioning results, including potential survey from both fixed reference electrodes and portable reference electrodes, current and voltage output values of each DC power source, calibration measurements for each fixed reference electrode and any adjustment made for nonautomatic devices

(9) The periodic maintenance inspection results, including reference electrode potential values, DC output values, maintenance data on transformer rectifiers and downtime periods; and

(10) An Operation and Maintenance Manual which details the as-built system, inspection and testing procedures, inspection, and testing intervals, and provides a fault-finding guide.

6.4 Combination System

A combination system is one containing both galvanic (sacrificial) anodes and an impressed current system. Records for a combination system should include complete information for each of the systems. As described in Chap. 3, Sect. 3.6.5, using a hybrid cathodic protection system (mixing SACP and ICCP) for turrets, sea chests, moorings, or risers, SACP potential may provide a potential level to cause ICCP system shut down.